T0131458

DENYING REALITY

*Comments on people whose view of the world
around them passes through a distorted lens.*

Pascal de Caprariis

authorHOUSE®

AuthorHouse™
1663 Liberty Drive
Bloomington, IN 47403
www.authorhouse.com
Phone: 1 (800) 839-8640

Published by AuthorHouse 07/09/2018

ISBN: 978-1-5462-5063-0 (sc)
ISBN: 978-1-5462-5062-3 (e)

Library of Congress Control Number: 2018908068

Print information available on the last page.

For Nancy

"Forever will I love and she be fair!"
Keats

CONTENTS

PREFACE

Here you will find a collection of short, related attempts to respond to a distressing development in today's society, namely the attitude that if you choose not to believe something, you are entitled to reject it, regardless of how little you may know about the subject. Unfortunately, that attitude does not seem outrageous to many people. In today's society, many people seem to take for granted that the difference between knowledge and ignorance is not important. How you feel is important, not whether you are correct, or whether what you believe conforms to reality.

As you make your way through this book, you will see that I reject that attitude. I was trained to believe statements that are logically sound and that are supported by evidence. Understanding how the world works is what is important to me. Facts, evidence, and logic provide information about reality, and I believe that everyone should be concerned about people who reject them. We should be wary of accepting the comments made by those who deny reality, because they have nothing useful to contribute to any conversation.

A classics scholar who published an English translation of the *Odyssey* anticipated criticisms of her work because she took some liberties with the way she told the story. She felt that if you think that a story is important, then it matters how you tell it, so she tried to find ways to express the thoughts in the poem in ways that a modern reader will be comfortable with. I took that maxim to heart in writing this book, because I suspected that few people would be fascinated by discussions of "facts" or "evidence," and I tried to keep in mind a comment made by the novelist William Golding about how easy it is for the reader to simply close a book and put it aside (He shuddered at the thought). So in this book I concentrated on thinking of examples of facts leading to evidence that support a claim. Surely, I felt, that with enough kinds of examples I can captivate readers who have a variety of interests, and show them that some of this supposedly dry stuff actually has meaning for them.

So the sections that follow are not extended essays on different aspects of how to think about reality, but relatively short ones that explain why I think that the ways some people do it are inadequate. Instead of thinking "this is what I choose to believe," people should be asking "what is the evidence for this statement?"

That question is a powerful tool, one which should be used by more people.

My point in this book is not to entice you to think about becoming a scientist; I just want to convince you that facts and evidence should provide the foundation of your understanding of reality. When you are comfortable with that way of thinking, your ability to get across the raging stream we call life will be easier because you will know where the rocks are as you make your way.

INTRODUCTION

"First lines are doors to the world."
Ursula Le Guin

Several years ago a book was published on Relativism, a philosophical position which holds that "absolutes" do not exist; that things we feel to be true are true only within our moral or intellectual framework.

The quote by Ursula Le Guin, fits the book you are holding perfectly, because everything you will find in it follows from the opening sentence. Everything in it pertains to the difference between people who believe in the reality of what we experience and those who for one reason or another, do their best to avoid thinking about it.

The book on Relativism contained as an example a discussion of a belief held by some Native Americans that their ancestors did not really cross over the Bering Strait from Asia, as claimed by archeologists who have carbon-dates from fire pits at different locations on the West Coast of Canada. Instead, the Zuni contend that their ancestors arose from within the earth once the spirits had prepared it in an appropriate manner. So, there are two inconsistent accounts here. Is the Zuni account as valid as that of the archaeologists? Relativists say that within their own contexts, both can be true, so there is no inconsistency.

But that requires that there be no independent reality.

Do Relativists really believe that? Do they understand that claims about the real world require some sort of supporting evidence if other people are to be asked to believe the claims? Is it reasonable to ask disinterested individuals to believe something just because you believe it, regardless of how much that belief differs from common knowledge?

The belief that the "spirits" were responsible for the ancestors of Native Americans does not differ significantly from the beliefs of other religions, but regardless of what their practitioners say, religious beliefs are ways to hold a culture together, not ways to describe reality. There are no independent sources which can provide unequivocal verification of the claims of different religions. How could there be?

In our society, much of what we know about the world in which we live comes from the observations we make as we go through our lives, but that source of knowledge is not entirely "absolute," because your experiences differ from mine, so we may well disagree on some things that we each have observed. But there will be a great deal of overlap. However, what is taught in school about the ways the world "works" is based on scientific studies that have been conducted since the period called the Enlightenment, in the 17th and 18th centuries.

What we call the "scientific method" represents a way of examining what is happening around us and analyzing what we find in those efforts. As new ways of thinking develop, and concomitantly, new technologies for doing the studies, we refine what has been learned in the past, and sometimes, reject it because it is not consistent with what we learn today. We learn things in this manner, but always remember that those things represent provisional knowledge, because with the development of better technologies and new understandings, some of what we "knew" in the past may be replaced. A distinguished Oncologist named Jerome Groopman once said that "science is the accretion of provisional certainties." We call them "certainties," because they satisfy the standards of the time, but they are "provisional," because we know that as new standards develop, some of those certainties will be superseded by new ones.

Although our technological society is based on the accumulation of knowledge determined in this manner, some people, who enjoy the benefits of the results of science, refuse to accept the findings of some studies. They do so for a variety of reasons, but whatever they claim is the reason, they are actually rejecting the existence of facts and evidence which are the basis of scientific studies. Implicitly, they refuse to accept that what we know about the world is based on facts. Real facts, not "alternate" facts and not hypothetical facts. Facts that are based on observations and measurements. Facts which are "reproducible," meaning that they can be verified by others.

In short, these people are rejecting reality. You cannot have a fruitful discussion with them.

But facts by themselves are not sufficient. They have to be compatible with basic principles of how the world works. Here is a quote from a distinguished scientist about what science is all about.

"Science is not a body of facts. Science is a method for deciding whether what we choose to believe has a basis in the laws of nature."

That statement was made by Marcia McNutt, a geophysicist who once was the director of the U.S. Geological Survey and, at the time of writing, is the President of the National Academy of Sciences.

Her point is that science provides a way to decide whether we should believe something. If the observations you make are compatible with basic principles, laws of nature, whatever we choose to call them, then they are reliable and are appropriate to use in further studies of the world. So, science progresses by making observations, evaluating them (are they compatible with laws of nature?), and putting them into an existing context (or creating a new one) called a theory. When the topic studied has been "vetted" in this way, it is given an imprimatur that satisfies people who believe in systematic approaches to examining reality. As was said above, when you reject the results of scientific studies you are rejecting reality.

When science says that something has a basis in the laws of nature, it is saying that the "something" is an aspect of the real world. A reliable aspect. One which has been subjected to independent verification.

What more can we ask?

Science provides a way to decide whether we should believe something. What other criteria exist that let us navigate the real world? Don't say "religion," because that tells you how to live your life, not how the natural world works.

Before moving on to discuss the meaning of "facts" and "evidence," I should clarify a few things.

Scientists know how to collect data. Reliable data. They know how to determine the accuracy of the measurements they make. They know how to analyze data using statistics to distinguish between meaningful relationships between two data sets and those that could have occurred by chance. And they know how to determine the uncertainty in the predictions they make.

Scientists are not perfect. Sometimes they make mistakes. But they are methodical and conscientious. And when they submit the results of their studies to a professional journal, it is examined by two or three experts, so the information published in the journals is about as reliable as we can expect.

So, when in subsequent pages I mention sets of numbers and relationships between them, you can safely assume that they were collected and analyzed by people who knew what they were doing. Just because some guy in a bar might discount a scientific result, does not mean that the result is worthless, because the guy in the bar is not likely to know what he is talking about, even when he is sober. And when a politician tries to avoid responding to a question about a scientific study by saying that "the scientists disagree," you can usually assume that he does not know how to answer the question. Most of the time, he is not likely to know how many scientists have contributed to the consensus opinion of the topics that are being discussed, nor will he know how few dissenters there are.

I will repeat the next few lines near the end of the book, because they are worth keeping in mind. In the 1920s, a sportswriter named Grantland Rice commented on how little sense it makes to bet against the favorite in a sporting event (that is, to bet against a consensus), by adapting a few lines from *Ecclesiastes*. He said:

"The race does not always go to the swift,
Nor the fight to the strong.
But that's the way to bet."

When you bet against the favorite, you usually lose your money.

Whenever you hear a talk or read an article about a scientific topic, you should pay attention to the "structure" of the presentation. Is it a bare-bones discussion of some conclusion scientists have come up with, or does it include the reasons for collecting the facts relevant to the topic, along with the logic behind how those facts are used to come to a conclusion? In the first example, the speaker is just telling you things, and those things may or may not be correct. How could you tell if all you hear are conclusions? In the second case, you are in the position of learning something, because the structure of the presentation allows you to evaluate whether it is sensible to accept the conclusions. You have facts and logic, so you can determine whether the conclusions are consistent with your previous knowledge of the subject, or not. In either case you have learned something. The time you spent listening or reading has been used fruitfully.

I hope you will feel that way when you finish this book.

FACTS AND EVIDENCE

Evidence is important, whether people want to believe that or not. And many do not. Even when a statement is supported by evidence, some people reject both the statement and the evidence supporting it because they do not want to abandon some cockamamie belief they have. This reluctance provides a nearly impenetrable barrier to conveying scientific information to the public, because it is very difficult to get people to examine critically some of their long-standing beliefs.

Let's start by distinguishing between facts and evidence. I will restrict the definition of facts to observations (which will include measurements), whereas evidence consists of facts that are relevant to some supposition. Facts are difficult to deny, at least for those who wish to appear rational. For example, if you tell people that the Sun will rise in the East in the morning, they will probably yawn, because that is a fact that everyone knows. And even if you tell them something that borders on the controversial, for example, that carbon emissions to the atmosphere continue to rise, most people will not react negatively, because by itself, the increase is does not threaten their pre-existing beliefs. So, that statement, can also be recognized as a fact. They will probably just say, "So what?"

On the other hand, consider this statement. "Most of the increase in the carbon dioxide concentration in the atmosphere in the last few decades has been due to the increased use of coal as a fuel source." That statement may be a fact, but it is not obviously one. It **is** a claim that requires some supporting evidence if it is to be useful in discussions of global warming and climate change. If you want a claim to be actionable, you have to provide some evidence for it.

Note that the statement, "the sun will rise in the east in the morning" is a fact, but it is not usually actionable. Unless your conversation with someone pertains to the manner in which the planet rotates about its axis, the statement does not require a response.

So there is a general rule for statements intended to be actionable. If you want to seem rational and credible, you need to provide some sort of evidence to support your claims. To take this point further, let's think about the source of the carbon emissions mentioned above. It turns out that evidence about the source of those emissions exists, and the evidence consists of facts, as it should. The facts involve the composition of the carbon in the carbon dioxide in the atmosphere. What, you may wonder, does "the composition of the carbon" mean? Carbon is carbon, isn't it? Not quite.

So, do I have your attention? Remember what I said at the end of the Introduction? About providing logic along with facts? Alright, I just made a claim about the source of carbon emissions. If I am to convince you that I know what I am talking about, I need to provide some evidence to support the claim. What follows is an example of how to use facts as evidence to support a claim.

The two main sources of carbon dioxide entering the atmosphere are combustion of fossil fuels (mostly coal and petroleum) and respiration, which refers to one of the processes by which humans and other animals get rid of waste products. That is, by exhaling. The carbon dioxide exhaled by us and by animals contains two stable forms (isotopes) of Carbon. The dominant one "weighs" 12 units and the other, much less common form, weighs 13 units (It is heavier because it has an extra neutron in its nucleus). The ratio of the two isotopes is constant across the animal kingdom, which is not surprising, because most animals use the same type of respiratory system to get rid of this gas as a waste product. Prior to the Industrial Revolution, respiration was the dominant source of carbon dioxide in the atmosphere. So those two processes, combustion and respiration, provide carbon to the atmosphere. What takes it out of the atmosphere? Plants.

We know that plants remove carbon dioxide from the atmosphere so they can use the carbon to make the proteins they need to function and grow. But they favor Carbon-12, not the slightly heavier one. The reason for that preference is that two enzymes used in their metabolic processes prefer the lighter isotope. So, over time, Carbon-12 is removed from the air preferentially, and the ratio of "13 to 12" in the atmosphere increases as the amount of Carbon-12 decreases (If the denominator decreases, the ratio should increase).

So, by determining the value of the ratio of "13 to 12" we should be able to determine the rate at which CO_2 is being withdrawn from the atmosphere.

Now, when we burn fossil fuels, such as coal, which formed from long-buried plant materials, we are burning a compound that is essentially pure carbon, and because it comes from plants, *nearly all of it is Carbon-12*. So, combustion of fossil fuels puts mostly Carbon-12 into the atmosphere, which should reduce the value of the ratio of "13 to 12" (because the denominator is increasing). And that is what is observed.

So one process involving plants removes the gas from the atmosphere and another, involving combustion of (former plants) puts it back in. The fact that the ratio of 13 to 12 is decreasing means that Carbon-12, the denominator of the ratio, is being added by combustion more rapidly than it is being removed by plants.

So, we have evidence that an appreciable amount of the carbon dioxide in the atmosphere is due to the combustion of fossil fuels. The evidence for that statement consists of facts. The first fact is that measurements of the amount of the carbon dioxide in the atmosphere show that it has been increasing over time. The second fact concerns the composition of that carbon, which shows that the increase is due to the combustion of fossil fuels.

A third fact that pertains to this subject is that carbon dioxide is a greenhouse gas, which traps some of the heat emitted by the planet. It follows that the increase in the planet's average temperature (that is, global warming) may well be due to human activities – unless some other mechanism is discovered.

Facts, plus logic, lead to a conclusion that is sound.

Evidence consists of facts that can be used to support a claim about some natural phenomenon. That is what the last few paragraphs do. When people discount the importance of evidence, they are rejecting the importance of facts. We should wonder on what they base their decisions if they are willing to play fast and loose with facts – which consist of observations. What does reality mean to such people? In what kind of world do they live if they can casually discount things that have been observed?

Does it make any sense to believe anything such people say?

One of the reasons some people refuse to accept conclusions based on evidence which indicates that human activities are causing global warming is self-interest. Many of the most vocal of these people are affiliated with the energy industries, whose profits depend on burning fossil fuels, which contributes to global warming. So their livelihoods depend on preventing any curtailment of their employers' activities. Others refuse to believe that warming is occurring due to carbon emissions is because they have been told that reducing those emissions to the atmosphere will devastate the economy, which makes them nervous about their future. No one wants to be depressed all the time.

Although it is true that large numbers of people are not willing to admit that human activities in the form of burning fossil fuels are causing climate change, it is also true that very few of those people know much, if anything, about the heat balance of the planet, so there is no reason to take any of them seriously, and there certainly is no reason to believe what they believe, just because many people choose to.

"Groupthink" is not a substitute for knowledge.

Remember, that although it is true that everyone has the right to an opinion about nearly any topic, it is *not* true that all of those opinions are equally credible. The people who have been studying climate for decades are bound to know more about the subject than some guy you might have overheard in a bar last night. Remember that truth is not neutral: truth is partial to knowledge. That bears repeating.

Truth is partial to knowledge.

Fortunately, you do not have to be an expert on atmospheric physics to understand why people who deny that global warming is occurring are wrong. All you need is a little information about how a gas "works," and some logic.

There I go again, combing facts and logic; just what I said is necessary if you are to "learn something" as opposed to memorizing what you read about this topic.

So, let's take the discussion to a slightly higher level by using some basic facts about the subject, along with the logic that converts them to evidence.

1. Carbon Dioxide (CO_2) consists of a carbon atom with two oxygen atoms located on either side of it. The molecule looks something like a dumb-bell. Now, a molecule is never still: it vibrates all the time, at a frequency determined by the relative weights of its atoms and the shape of the molecule.

2. Chemists have measured the frequency at which CO_2 vibrates, and found that it is in the range of what we call infra-red (or heat) radiation. That is, as heat radiation leaves the surface of the planet in the form of waves, it is oscillating at about the natural frequency of CO_2. And because of its structure, the molecule is sensitive to that radiation.

Both of those statements are facts. They have been observed in carefully controlled laboratory studies.

3. As the heat radiation passes through the atmosphere, the CO_2 molecules there are "tweaked," and vibrate more vigorously. This effect has also been measured in the laboratory. Tests involving sending heat radiation through a transparent box containing nothing but CO_2 have been done. The intensity of the radiation coming through one side of the box is less than that entering it on the other side, and the temperature of the gas in the box increases. So, the gas is absorbing some of the heat. That is a third fact.

Those three statements are facts, not opinions. I call these statements facts because they are based on measurements. Numbers. The output of instruments. Combined with some basic physics, which has been known for more than one-hundred years, we end up with the kind of conclusions published in the U.N. climate reports. The increase in the planet's average temperature is occurring along with an increase in greenhouse gases in the atmosphere. The conclusion drawn from this correlation is that the emissions are causing the rise in temperature of the gases that make up our atmosphere. So the claim often made by "climate-change deniers," that the correlation between the increase in CO_2 and the increase in temperature does not necessarily mean that one causes the other, is just a subterfuge. There is no credible alternative explanation for what is observed.

The statement that the conclusion is consistent with the basic principles of physics reflects Dr. McNutt's definition of science, in the Introduction. The conclusion is worthy of belief.

Any rational person has to realize that Global Warming is occurring, and that carbon emissions from the burning of fossil fuels is the only reasonable cause.

Unless someone has measurements of something that contradict the facts listed above, and also some physical principles that contradict basic principles of physics, these facts and the conclusions drawn from them are meaningful, believable, and important.

It should be easy to convince people of all this, but in fact, it is not. The basic level of scientific literacy of the general public is low, so a discussion based on atmospheric physics, even at a very low level, is easy to dismiss. In addition, the tendency of people to ignore or reject any recommendations that would force them to change the way they live is very strong, therefore a substantial number of people do not take seriously the warnings climate scientists have been issuing for decades.

But just in case the logic behind the warnings is strong enough to have an effect, the energy industry has been funding efforts to convince people that they need not worry about what the industry claims to be the relatively minor effects of burning fossil fuels.

If the coal and the oil and gas industries were to leave these materials in the ground, society might avoid the catastrophic consequences of global warming, but that option would destroy those industries. The companies would have to shut down, because their assets are worthless if they are left in the ground. Needless to say, that is not an attractive option for the people who make a very good living by extracting and burning these assets.

The test with the transparent box mentioned above indicates that the energy used to cause the increase in the vibrations of the CO_2 molecules in the atmosphere stays with those molecules for a while, so it does not go out to space immediately. That is a basic application of logic. If some of the energy causes a molecule to vibrate more vigorously, it cannot be going out to space.

That statement is a way to describe a basic principle of physics, namely that "Energy is Conserved." That is, energy can change from one form to another, but it cannot "appear" from nothing or vanish. If its value changes, it has shifted from one form to another, e.g. from radiation to molecular vibrations, vibrations which are equivalent to heat.

The overall conclusion drawn from these facts is that the shape and composition of the CO_2 molecules keep some heat close to the planet's surface longer than if those molecules were not in the atmosphere. That is the basis for the phenomenon called the Greenhouse Effect, which keeps the surface of the planet about $60°$ F warmer than it would be without CO_2 and other gases with somewhat similar properties.

The totality of the facts provides the evidence. No amount of dissembling (or deception, guile, misrepresentation, delusions, or whatever one wishes to call the statements of "climate-change deniers") can change that.

You do not have to be an expert in climate science to understand this. Just remember that CO_2 keeps heat close to the ground surface for a while.

The next two sentences are very important.

Based on the facts in the last few paragraphs, can anyone doubt that if we increase the amount of CO_2, more heat will be trapped? If it is not trapping heat, what is the gas doing?

That is a question the "climate-change deniers" never try to answer, because they cannot. What is the gas doing?

I should add that carbon dioxide is not the only greenhouse gas. Methane, Nitrous Oxide, Ozone, and Water Vapor also trap heat in the atmosphere. And some of the chemicals that were once used in air-conditioning systems also trap heat, but with the exception of methane, none of the gases other than CO_2 are increasing in the atmosphere. Methane is an exception, because some is released from gas drilling operations, but at present, the amounts released are small. In addition, the flatulent emissions from livestock contain methane, but unless livestock production increases markedly, CO_2 will continue to get all the attention because it is the only one over which we have any real control.

But - and this is a large "but" - a lot of methane is trapped in soils in the permafrost regions in Arctic areas. As the planet continues to warm, these soils will thaw and release that methane, which will cause global warming to accelerate.

In this chapter, I hope to have convinced you of two things. First, that facts and evidence are not entirely boring topics, and second, that fact-based evidence supports the statement that human activities are responsible for the observed increase in CO_2 in the atmosphere and the rise in the average temperature. With those things behind us, let's go on and think about a different application of the discussion about facts and evidence.

MAKING ASSUMPTIONS

Scientists are often irritated by the casual attitude people have toward scientific studies, and a major source of irritation is how people are willing to discount the results of a study because one or more assumptions have been made in setting up the study or analyzing the data collected. People do not understand that assumptions are tools that allow us to examine a phenomenon. If the assumption is valid, we may be able to develop a better understanding of the phenomenon. If the assumption is not valid, we usually end up at a dead end, and the assumption is discarded. So it is not *just* a tool; an assumption can be a very useful tool.

Let's examine an example by discussing how we determine that the age of the planet is 4.54 billion years, something contested by people called creationists, a group of biblical literalists who point out that the Bible implies that the Earth is only about 6,000 years old, so of course they reject any scientific study that might contradict that belief.

What is the evidence that the Earth is 4.54 billion years old? It comes from measurements of the rate of decay of three radioactive elements, of which Uranium-238 is the best known. That form of

Uranium decays to Lead-206 (instead of the much more common Lead-208). Lead-206 only forms from the decay of Uranium-238, so if you measure how much of each of those two elements are in a mineral, you can use the rate at which the transition occurs to determine how long it has been occurring. That will be the "age" of that mineral and the rock that contains it.

That seems simple enough, but let's repeat it for emphasis. Assume that a mineral in a rock contains an element A, which undergoes a transition to element B, and does so at a known rate R (say, at so many micrograms per year, for convenience). If you measure the amount of B (the new one) and then divide it by the rate R, you are dividing micrograms by micrograms per year, so you end up with years, which represents the length of time it took for that amount of B to form. That is how long the transition has been going on, and it is the "age" of the mineral containing A, and presumably, the age of the rock containing that mineral.

Note what is involved here. Measurements of the amounts of the two elements can be done precisely. The values are "facts." The measurement of the rate at which one decays to the other can also be done precisely. That is another fact. What remains is the logic (dividing an amount by the amount that changes per year). That is the level of algebra taught in middle-school math classes, and combined with the facts mentioned, is the basis for what is called "radioactive dating" techniques. The measurements (facts) and the logic provide the ability to determine the age of the planet. The value obtained is 4.54 billion years.

That last paragraph is another example of what I said in the Introduction about the importance of giving you the logic behind a statement. When you have all of the relevant information and logic, you can evaluate a statement; there is no need to accept anything just because the person presenting it seems to know what is going on.

Some people, the ones called creationists, contest everything in this argument. They claim that the Earth cannot be billions of years old, because the Bible clearly states that it and everything on it were created once, in six days, about 6,000 years ago. It seems that you get a date of about 4,004 BC for the "creation" if you add up the ages of the men in the Old Testament. Of course, that requires you to believe that every story in the Old Testament is literally true. Even the stories about men living for several hundred years. That is quite an assumption. But the creationists don't consider that an assumption. More on that later.

They do their best to discredit the scientific studies by claiming that the age obtained from radioactive dating techniques is not reliable because it is based on an assumption that cannot be proved, namely that the decay rate has not changed in the past. They go on to say that if the technique used to get an "old" value for the planet's age is not reliable, then the Biblical account *must* be true.

Their argument is basically, that if the reliability of radioactive dating techniques is at all questionable, we should reject this approach out of hand, in which case the only reliable estimate of the planet's age is that found in the Bible, which they assume is inerrant. So in addition to assuming that every story in the Bible is literally true, they are assuming that the Biblical account is the *only* alternative if the scientific results are not reliable. But they don't mention why their two assumptions are reliable. They want us to reject science and accept Bronze Age mythology. And they expect to be taken seriously.

It *is* true that if the decay rate were not constant, the radioactive dating technique would not be reliable, but the planet could *still* be billions of years old, even if the "clock" used is inaccurate. However, there are a variety of ways to justify the reliability of the radioactively-determined ages.

Let's simplify the discussion a bit. It is possible to measure accurately the rate at which Uranium-238 decays to Lead-206. The rate is very slow, and to give a feeling for how slow, it is customary to refer to the "half-life" of Uranium. The half-life is the number of years for one half of the amount you start with to transition to Lead at the rate of decay that we measure. The figure that has been determined is 4.468 billion years. Let's say 4.5 billion (It is a coincidence that the value is approximately the same as the age of the earth). Using that figure, determination of the age of the planet is made from meteorites, whose ages are used as proxies for the age of the planets in our solar system, and the results give the value of 4.54 billion years. We use meteorites because as fragments left over when the planets formed, they have not been subjected to geologic forces (both physical and chemical) that have acted on planetary materials since the planets formed. So, they are pristine.

Creationists contend that regardless of the value obtained for the half-life, no one can prove that it has not changed in the past. After all, no one was around a thousand, much less a billion years ago to measure it the decay rate, so who can say that it always has had that value?

That is the argument they make to cast doubt on the techniques of radioactive age determinations, and which they use to claim that their belief in biblical inerrancy should be accepted as more reliable. And that is the point of the discussion that follows, in which I show why the decay rate and the half-life of a radioactive element cannot change. Not that decay rates *have not* changed, but that they *cannot* change.

But before going into any details, let's examine what creationists consider to be a strong argument. They contend that if you were not there when something happened you cannot be sure what happened. If no one was around a billion years ago, no one can demonstrate that the decay rate (or half-life) was the same as that measured today. So, they claim that we have to have been there in the past before we can consider the age reliable. But, do we?

In her book *The Canon*, the science writer Natalie Angier mentioned this argument, and used the following example to show how absurd it is. Nearly every week, somewhere in America, someone is on trial for a serious felony. Forensic experts who examined the scene of the crime testify at the trial and present what they found. Can you imagine the reaction if a defense attorney told a forensic scientist that his interpretation of what happened is not credible because he was not actually at the scene when the crime occurred? That attorney would be laughed out of the courtroom.

In that example, as in all scientific studies, evidence counts. If evidence is collected according to prevailing standards and the data are analyzed using the best techniques, no one who wishes to seem credible is going to contest the conclusions just because the experts were not there when the crime was committed. The point of the example is to suggest that there are reliable ways to determine some things that happened in the distant past.

So let's examine the credibility of claims that the Earth is very old by evaluating the assumption that decay rates of radioactive elements have never changed.

To give you a "heads up," I am going to argue that what the creationists criticize as an assumption is not at all speculative. I am going to present evidence that it is a fact, in which case, the argument of the creationists is specious.

Imagine that you are holding a small piece of Uranium-238 in your hand. That would not be a good idea, because it *is* radioactive, you know. But imagine you are. If the piece is a cube, about one centimeter on a side (about the size of the fingernail on your pinky), there will be a very large number of Uranium atoms in it. A rough estimate would be one times ten to the twenty-fourth power. A trillion is one times ten to the twelfth power, so that cube has about a trillion squared atoms.

Each one of those atoms is unstable and eventually will decay to Lead-206. We cannot determine which of those atoms will decay at any given time, because radioactive decay is a random phenomenon, but we do know that all of them have the potential to do so. And the best measurements made in the last one hundred years indicate that in 4.5 billion years, half of the trillion, trillion atoms will have changed to Lead.

The question is, how do we know that that value has never changed?

To answer that question, we need to know what "decay" means. The nucleus of a Uranium-238 atom has 92 particles called "protons," each of which has a positive electric charge, and 146 "neutrons," which have no charge. The sum of the two numbers gives 238, the Atomic Weight of the atom. Because they have the same positive charge, protons tend to repel each other (remember, "opposites attract and likes repel"). Crowded in the extremely small volume of the nucleus (about a million times smaller than the diameter of the atom itself), the force of repulsion between the protons is very strong, and that should cause all of them to fly out of the nucleus and cause it to vanish. But the job of the neutrons is to shield the protons from each other to prevent that from happening. The problem is that in the "heavy elements, there often are not enough neutrons to do the job properly, so eventually, a clump of particles (two protons and two neutrons) is kicked out of the nucleus. It comes out as radiation, because according to Einstein's equation $E = Mc^2$, the mass, M of that clump (multiplied by the speed of light squared) is equivalent to a certain amount of energy E. We can measure that radiation with a Geiger Counter. When the clump is ejected, the Uranium atom "decays" to a different element, which itself, decays, and so on. In about three or four steps, the result is Lead-206.

Note: The description in the last paragraph is how the process is described in a high school Physics course (such as the one I taught in an earlier life). The word "shield" is a euphemism for what the neutrons do, but to understand what actually happens, we would have to know what protons and neutrons are made of, so we could understand how they interact with each other. That would require course work in nuclear physics (which I never took), so we have to be content with the word "shield." End of digression.

Getting back to the point, now that you know what is going on, imagine another (very small) piece of Uranium-238 - small enough that it consists of about 100 atoms. (You would not be able to see something that small, but imagine that it exists.) There is no way to predict which of its atoms will decay, or when any of them will. Because they all act independently of each other, in principle, they could all decay at one time, or at equally spaced intervals, or at random intervals. And some of those intervals could be longer than our lifetime, in which case, we would never know when any of the atoms decayed.

In this example, the timing of the decay process is not predictable because there are so few of them *and* the decay is a random process.

Now consider a larger piece, one with one thousand atoms. Nothing has changed: there are still too few for us to expect predictable behavior. How about a million atoms? Or a billion? Or a trillion? Well, when we have a very large number of atoms, things change.

When the number is very, very large, some of the atoms will be decaying all the time. The time intervals between the decays of the original 100 atoms are filled in by the decays of the additional ones, and when there are so many of them, what we observe is a regular process. *But only when the number of atoms is very, very large.*

Using a Geiger Counter, we can monitor the bursts of radiation, determine how many have decayed in some time interval, say a year, and knowing how many we started with, we can determine the time it will take for half of them to change to Lead. That is, we can determine the half-life.

But we can do that *only* when we have an enormous number of atoms in the sample, because until the number is enormous, the behavior will not be regular enough for the concept of decay rate to be meaningful.

Do you see the point? When large numbers of the atoms are present, a new property becomes meaningful: the decay rate. That property does not exist unless a very large number of atoms are present. That principle applies to many things. Think of H_2O molecules. When you have a relatively few, say a dozen, they exist as a gas, floating around in the air as water vapor, but when there are an enormous number of them, they "glom together" and change to the phase we call liquid water, at which point, new properties emerge, ones which could not be predicted from their behavior as a gas. For example, nothing will dissolve in just a few water molecules in air, so the property we call solubility makes no sense when you are dealing with water vapor, but the concept of solubility does make sense when you have a few trillion molecules and they are in the phase we call liquid water.

Getting back to Uranium-238, if the property that we call the decay rate, or half-life "emerges" (that is an important word) only when we have very large numbers of atoms in a sample, then that property *cannot be a property of any single atom*. It is a collective property of the entire collection or ensemble of atoms, and as such it either exists (when the number of atoms is large) or does not exist (when the number is small). But when it exists there is no way for it to change. There is no physical mechanism that would cause the rate of emissions recorded from the finite sized piece of Uranium in your hand to change because that rate is not due to anything within the nuclei of the atoms. No atom's behavior affects the behavior of its neighbors. They behave independently.

"Emergent" properties such as solubility and decay rates are constants. They are properties of the collection of particles, so the decay rates of radioactive elements such as Uranium-238 are constant. They have not changed in the past because they *cannot* change.

I warned you that I was going to present evidence that the decay rate or half-life of an unstable element cannot change. The consequence of that result is that it is appropriate to call that rate a fact, and consider it to be just a normal component of the logic used to determine the age of a mineral containing a radioactive element. And I think that anyone except a creationist could follow the logic presented in this section.

Assumptions do not necessarily weaken an argument, and certainly not when one can show that they are facts. And facts that support a supposition represent evidence that the supposition is true. The value of the decay rate is a fact. It follows that based on radioactive age determinations, the 4.5 billion-year age of the Earth is credible.

Although I give very few references in this book, I feel I should mention that I found the concept of "emergent" processes in a book (*A Different Universe*) by Robert Laughlin, in which the concept is applied to a number of topics in Condensed Matter Physics (nee Solid State Physics). Radioactive decay is not one of them, but the application was so obvious that I used his logic to develop the argument presented here.

An Economic Example

Now let's consider an assumption that can be said to be speculative. Not necessarily wrong, but one which must be justified. Some of the consequences of global warming involve changes in climate, but the precise effects of the changes on our society depend on how certain calculations are made. For example, politicians and Economists tell us that if society curtails the use of fossil fuels, the world's economy will be devastated, so until scientists can "prove" that climate change will happen, and that it will be as drastic as they claim, we should not make any substantial changes in how we use coal and petroleum, which "power" our economy.

So, we need to estimate what the precise costs of addressing global warming will be, and compare them with the costs due to significant changes in the climate if we do nothing.

In either case, a number of assumptions must be made, because after all, we are talking about predicting the future. The most important of those assumptions involves what Economists call the "discount rate," which is an estimate of how much a dollar spent today would buy in (say) ten or twenty or more years, due to inflation. When you choose a value, you can estimate by how much the economy will differ in some specified number of years from today's behavior.

Notice that the value of the discount rate you choose depends on how you think inflation will affect the prices of things at some specified time in the future. It truly is an assumption.

Nicolas Stern is a British Economist who supervised an extensive study about the economics of climate change. The report he wrote is structured around the Precautionary Principle. Note how the word contains its definition. Pre-caution. Be cautious before doing something. In the study he supervised, he noted that there are two kinds of mistakes we can make with regard to the information we have about climate change.

We can act on the scientific advice only to find out that it turns out to be wrong, or we can assume that the science is wrong, and do nothing, only to find out that the science is right. Which mistake, he asks, is more dangerous?

If we act and find that it was not necessary, we benefit because we will have a cleaner world, and the overall cost of running it will be less than today's cost. There would be costs associated with that option, of course, because we would have to spend a lot to substitute alternative sources of energy for the fossil fuels, but his study concluded that those costs, although substantial, would not be disastrous. In fact, Stern concluded that the cost to society of curtailing the use of fossil fuels would be less than the cost of dealing with the severe effects of climate change if we do nothing. So, he concluded that it is cheaper to do something than to do nothing.

Of course, his cost analysis depended on the assumption he made about the discount rate. Stern's conclusion that addressing climate change was the appropriate policy generated controversy, because it contradicted the litany from politicians, who are always averse to major changes, and from the usual group of climate-change deniers, who would have us believe that curtailing the use of fossil fuels would destroy the economy.

Stern's detractors did not question the use of a discount rate, just the value he used, because it produced results they did not like. They did not quibble over the use of an assumption; they quibbled over the value assigned to the assumption. The assumption was a tool that was needed to get an answer to a question. His detractors felt that he got the wrong answer, so they decided that he must have used the wrong value.

How can those of us (myself included) who are not professional Economists decide which side of this controversy makes more sense? One way is to examine the subtext of the controversy. The critics of Stern's report are unhappy because he concluded that the effect of reducing emissions to prevent disaster would be a reduction in the global Gross Domestic Product of about one percent. For comparison, he said that doing nothing and letting climate change accelerate, would result in a drop in the global GDP of between 5 and 20 percent.

What does that say about the detractors? One obvious thing seems to be that they consider increased economic growth to be sacrosanct. Anything that interferes with that goal is unthinkable to them.

One percent versus between five and twenty percent. A relatively small cost versus a potentially devastating cost. Who could complain about a relatively small cost, considering the implications of the other possibility? The answer is, those who are convinced that the only thing that is important is economic growth.

Here is a bit of logic.

It seems clear to me that Stern's critics consider economic growth to be the *end*, rather than the *means to an end*. To them, a functioning society (the *end* or goal for most people) is not as important as ever increasing economic growth. They don't *see* the devastated society that will be caused by significant changes in the climate. All they *see* is a lower value for the economic growth rate. They are wearing blinders.

The details of Stern's analysis can be found in his 2007 publication. If you are willing to read about economic modeling, do a search on *The Stern Review*. Stern also published a volume for the layman, titled "*The Global Deal*," in 2009.

In this discussion, I've given two examples of how critics of a study use the concept of an assumption to downplay (or completely discredit) the results of a study that they are not comfortable with. When critics of technical reports cannot show that the reports are flawed, they resort to diversions (such as criticizing Stern's choice of the discount rate). A little sleight-of-hand often is enough to distract people from considering whether policies based on the conclusions of technical reports are worth implementing.

I will discuss this kind of diversion again in the last chapter, but for now I will just say that Stern's critics have an agenda to promote, regardless of its compatibility with reality.

Hoist Them by Their Own Petards

Now let's examine another example of assumptions by considering some that creationists use, even though they cannot be verified or proved or shown to be factual - however we want to describe the criterion - and think about the validity of the conclusions that are based on those assumptions.

Creationists would have us assume that the Old Testament is inerrant; that the stories told in it (creation in six days, a worldwide flood, the Sun orbiting the Earth, etc.) are literally true. Based on that assumption, they contend that theirs is the only "true" religion and that educational curricula should be modified to accommodate their beliefs. The main belief they want incorporated involves the Theory of Evolution, which they say should be scrapped in Biology courses, because, they say, the world, including the human body, is designed so well that there had to be a "designer." They go on to say that because the biblical story of the creation "dates" it as occurring in six days, about 6,000 years ago, it should be clear that random mutations of DNA molecules could not possibly have resulted in what we see around us, because there would not have been enough time for evolution to happen.

So, let's consider the assumption of inerrancy. But first, let's create a context for the discussion.

Imagine that a fellow goes to the manager of a television station and claims to be receiving radio signals from extraterrestrials through the fillings in his teeth. He says he wants air time to tell the world what the aliens told him.

Would any sensible manager give him air time? Of course not, because the fellow clearly is a lunatic. But if the manager chooses to be diplomatic (instead of just throwing the guy out of the office), he might ask the fellow to answer a simple question.

"If it is possible to receive radio signals through the fillings of one's teeth, why don't I pick up the signals through the fillings in my teeth?"

This is the sort of question that would have to be addressed to establish the fellow's credibility. If you want me to believe what you are saying, you have to provide evidence, and that evidence should be based on facts.

This example is not entirely far-fetched, because it is analogous to claims made by biblical literalists. They make claims that are not consistent with scientific findings, and which have no supporting evidence. They ask us all to accept their claims, which they say are consistent with the material in the Old Testament, and which therefore must be true. A religious book written in the seventh century BCE based on Bronze Age mythology is the source of the evidence they use to reject modern scientific studies.

So, the "facts" that support their arguments come from the Old Testament. And on what is the credibility of these "facts" based? Well, divine revelation provided to those who wrote the book is the basis for believing these "facts."

Is it obvious that this line of "reasoning" is essentially the same as that used by the fellow with the sensitive teeth? Apparently not, because no one in his right mind would believe comments about communicating with aliens, but large groups of people in America have no difficulty in believing that divine revelation gives credence to Bronze Age myths which therefore are more reliable than modern scientific studies.

The problem creationists must address is that although there is no way to disprove the existence of divine revelation, there is no way to prove that it exists. The "proof" they provide is to say that revelation clearly is found in the stories in the Bible (For example, God talking to Moses on the mountain). So we are asked to accept circular reasoning. The stories in the Bible must be true because revelation is their source, and we know that revelation is credible because it is found in the Bible.

These are not facts: they are at best, hypotheses, but ones which have never been demonstrated to be valid. If creationists want us to believe that they are acting the way scientists act, they should be able to answer the following questions:

What evidence is there that the earth is only about 6,000 years old? That is, prove that the earth is young, *without making the assumption* of divine revelation.

What evidence is there that radioactive dating techniques are not valid? Prove that the half-lives of radioactive elements are not constant, and do that without making any assumptions.

Of course, to do that they would have to develop a theory of nuclear processes which showed that atoms acting independently of each other interact and change the behavior of their neighbors. If the inconsistency in that last sentence is clear to you, congratulations: you are paying attention.

If creationists could do such things, they would have done so by now. They rely on circular logic to "prove" their claims, using arguments they would condemn if used by scientists, but they do not seem to realize what they are doing. They always have an "escape hatch," in the form of "it must be true because it is in the Bible."

That is not how scientists work. There is no reason to accept any of the arguments creationists make because, ironically, their assumptions are not valid.

Creationists would be harmless if they confined themselves to insisting that their religious beliefs were true, just as the proponents of every other religion assert about theirs. But creationists persist in denying that the Earth is billions of years old because they need everyone to believe that the Earth is only about 6,000 years old. The reason is that that short interval of time would not be enough for evolutionary changes to occur from one-celled organisms to animals, to primates, to human beings. That series of changes contradicts the "creation story" in the Old Testament, which they believe is literally true. So, if evolution could not have occurred, they feel that their religious beliefs are the only alternative and represent what should be taught in public school Biology classes. That would violate the First Amendment, of course, but that is a minor matter to creationists, because if their account is "true," they feel that it would be "science," and therefore would not present a Constitutional problem.

So creationists want a religious belief, for which there is no independent verification (and for which there cannot be any independent verification) taught instead of the body of knowledge involving science which has been established on the basis of facts and logic in the manner of the material discussed in the second chapter.

I repeat: That is not how scientists work. There is no *reason* to accept any of their arguments because they are not valid.

APPROPRIATE USE OF LAND

The words fact and evidence seem straightforward to me, even though some people find them difficult, or at least ambiguous. If you got this far in this book, you know that I think that those people have a tenuous grasp on reality. But those two words are not the only important ones that seem ambiguous to some people. The word "appropriate" is another one that some people have problems with.

"Appropriate Use" is a term that should be recognized by everyone as just common sense. After all, who would approve of an "inappropriate" use? But "appropriate" differs from "fact" and "evidence," because it is an adjective rather than a noun, so it is a label, and therefore it involves a value judgement. When you tell a man that "the facts do not support your argument," you are opening a discussion that involves logic. But when you tell him that "your plan is not appropriate," you are criticizing his judgement. You are telling him that what he wants to do is wrong. The difference is not subtle. Saying that his logic is faulty is not equivalent to invoking moral reasons to criticize his plan.

When you are using adjectives, you have to be careful. Sometimes, people react negatively to value judgements

When geologists or engineers or plan commission members use the words "appropriate use," they usually are thinking in terms of facts that involve the characteristics of a piece of property and matching them against the plans a landowner has for it. Those facts might involve the soil type, or steepness of slope, or the propensity to flood, or the depth to the water table, or any number of characteristics of a piece of property. Evaluating a proposed use for the property involves determining how the land would "respond" to the proposed use. That determines whether the use is deemed to be appropriate or inappropriate.

The word "respond" is important. You can use a table for any number of things. You can serve a meal on it, write a letter on it, stack a pile of books on it, or you can even stand on it to paint the ceiling. The list could go on and on. The table is an inert object: it does not "respond" to the way or ways you use it. *But land is not inert.* And even though you may not realize it, lots of things are going on within the piece of land you would like to use in some way. And, like it or not, some of those things may cause a response to your proposed use in ways that can impose costs on you, your neighbors, or on the entire community.

So, value judgements are involved with the use of the word "appropriate," and those judgements reflect the desire to prevent consequences that may impose costs on the community if inappropriate uses are conducted. As I asked before, who would want to choose something that is inappropriate? The only way to justify such a choice is to reject the science behind the determination of what is likely to happen if such a choice is approved. Like it or not, to contest those judgements is to reject facts. People who reject facts are going through life wearing blinders.

Consider the example of flood plains, which are the flat areas bordering large rivers. They have always been attractive sites for development because they are flat, which makes construction relatively simple, and they have a river flowing through them, which in the past, facilitated transportation. But here are two important facts about flood plains.

First, those flat areas exist because the river occasionally overflows its banks and covers the area around it with water and sediment.

Second, when that happens, the water continues to flow downstream and the sediment suspended in it settles and puts another layer of soil on the area adjacent to the river.

These two facts are awfully hard to deny. Anyone who lives near a large river knows that these things happen.

Now let's consider some consequences. Consider the second fact given above.

"When that happens, the water continues to flow downstream..." That is important because it indicates that *the flood plain is really part of the river bottom.* The sediment that is suspended in the flowing water settles out and covers the area around the river with new material. So the thickness of the soil in that flat area increases. That flat area was created by the river, first by erosion of the surrounding area and then by accretion of new soil from what is suspended in flood waters. It was created by the river, and from time to time it serves the same purpose that the river channel serves - it allows water to move downstream.

We forget that it is part of the river because the river does not need to use it very often. So, in a sense, we borrow it from the river for our uses. But when too much rainfall occurs upstream, the volume of water coming down the river is greater than can be transported by the normal river channel, and the excess water flows over the bank, temporarily reclaiming the area it normally does not need, and uses the expanded area to transport the excess water downstream. In this way, the size of the river adapts to changing circumstances. In one way or another, the river is going to move that water downstream and send it on its way to the ocean.

So, the fact to keep in mind is that the flood plain is part of the river. The evidence for that claim consists of the two facts given above, which are familiar to anyone who has lived on or near a flood plain for an extended period of time. In addition, the layers of sediment in the flood plain usually differ from the soils in the area surrounding the flood plain, because they have been carried there from "upstream" in the watershed. How else could they have gotten there if not by means of river transport? And what else could cause them to be found reasonably far from the river's edge, if the river itself had not "expanded" and put the sediments where we find them? In addition, the "flatness" of the plain often contrasts with the topography of the surrounding area, which is what you would expect from an area built up by flood waters.

It is a fact that the flood plain is part of the river, because it was created by the river.

The last sentence implies that *we must be very careful* in "developing" the area surrounding a large river.

The flooding in Houston from hurricane Harvey, and some observations from a visit to a small community in North Carolina at about the same time provide examples of how the concept of appropriate use in flood plains affects our ability to "develop" these areas.

The flooding in Houston from hurricane Harvey in 2017 was severe because of inappropriate use. For generations, development was sanctioned in and near low-lying areas that are prone to flooding after heavy rains. The hurricane was an extreme example of what can happen, dumping a year's worth of rain on the area in a few days, so some flooding due to the hurricane would have occurred even if development in low-lying areas had been restricted. But in fact, Houston and the areas around it in Harris County have few restrictions on building, so development in low-lying areas was rampant, and the damage from the hurricane was extreme.

At this point, some details about flood levels seem appropriate. The "100-year flood" refers to water reaching a specific height above normal flow levels, and the probability that it will reach that level is one percent in each year. So in principle, there is a one in one hundred chance that the flood waters will reach that level in any given year.

But, and this is a substantial "but," the longer the interval between 100-year floods in a particular area, the greater the likelihood that it will occur *this* year. That is important. The longer between floods of a certain level, the greater the chance that one will occur in any given year.

For example, if a 100-year flood has not occurred in an area for 30 years, there is a 26% chance that it will occur this year (not 1%). If the interval is 50 years, the chances rise to 39%.

Those numbers should be enough to convince you that when realtors reassure you that the houses or lots they are selling (which are in the 100-year flood plain) have not been flooded for many years, you should walk away from the sale. If there have been enough floods in the area for the elevation of the 100-year flood level to be determined, there are enough data to make predictions of the likelihood of future events.

If you are curious about the numbers given above, go to Wikipedia and search on "one hundred-year flood." You will find a formula from which you can get the numbers above by using a calculator. It may look formidable, but all that is needed to get numbers out of the formula is division, subtraction, and then raising the result to a power. For example, if you insert 100 in one place and 30 in another, the formula will give you 0.26, or 26% for the likelihood of a flood if one has not occurred there for the past 30 years.

Most of the properties with flood damage due to Hurricane Harvey were within the 100-year flood plain, but the homeowners did not know it. The reason many homeowners did not have flood insurance is because for several years, developers trucked in tons of soil to raise some areas an inch or two above the 100-year flood level, and had the flood plain maps revised. The properties were then sold to people who were told they did not need flood insurance because the properties were above the critical level.

One more fact. Even when the maps are not modified for "reasons of development," the elevations given on the 100-year floodplain maps are not usually as accurate as one might wish, so some properties may escape flooding even though on the map, they seem to be in the "danger zone" for a 100-year flood, and others that appear to be a bit higher than the 100-year elevation may be flooded. Most of the time, prospective home-owners would be wise to avoid property that is in or close to zones that require flood insurance.

Knowledge about the dangers inherent in developing flood plains is not tucked away in professional journals that only experts read. Anyone with a grain of common sense knows that areas that have been flooded before will be flooded again. But the powers-that-be in the Houston area seem to have been more interested in the economic potential of some parts of the area (as long as they are dry) than with worrying about the economic costs of dealing with the aftermath of floods.

They ignored facts and evidence.

Now consider another example. In the summer of 2017 my wife and I took a week off and visited Beaufort, in Eastern North Carolina. One day, we drove through the small community named "Oriental," which is a about a half hour's drive from Beaufort, a drive that was very instructive. The community is adjacent to a large river that flows into Pamlico Sound, which is protected from the ocean by the Outer Banks. Oriental is in the flood plain of that river and so is subject to periodic flooding. That was indicated when we drove through a cluster of houses, not adjacent to the river, but within sight of it, because every house was raised on stilts, about four or five feet above the ground. One house, which has a porch in front, has lattice-work covering the area beneath the porch. From the ground level up to about two feet, the lattice-work had water stains.

The fact that every house in that subdivision was raised on stilts indicates that the local building codes reflect the fact that parts of the community have been subjected to floods. The building code reflects reality. Apparently, unlike in Houston, you cannot build in a flood plain in Pamlico County, North Carolina without taking that reality into account. The difference between Oriental and the Houston area is pronounced. Even after three significant floods in three years, the political climate in Houston is such that significant changes may still be difficult to be incorporated into the building code.

You may wonder why would anyone want to build so close to a major river if the flood plain is part of the river. That is, why would anyone build on land that evidence shows is going to be flooded again at some unknown time in the future?

Consider an equivalent question: Why would society continue to burn fossil fuels when evidence shows that it is causing climatic changes which will be exacerbated with time if the practice continues? Is that activity "appropriate?"

Perhaps the answer to both questions is that people do not want their plans to be thwarted by the possibility that some sort of disaster will occur in the future because of those plans. Sometimes they end up putting restrictions on development, but in other places, they ignore the potential for the problems, or they deny that the problems are significant, or they deny the very existence of the problems. Perhaps they hope that someone else in the distant future will be the one to "pay the piper."

Unfortunately, the "distant future" is not always distant. That is why we have zoning regulations in floodplains.

How many events such as hurricane Harvey will it take to get the people in Texas to recognize reality?

Some people claim that they have the right to do anything they wish on property they own, so the object when a local government agency tells them they cannot build at some location on their property. They feel that ownership of property is absolute and not subject to any local ordinances. But they do not always feel that way about other kinds of ordinances, laws, or regulations.

We routinely accept many kinds of regulations that restrict our right to do certain things, so there seems to be no reason to believe that rights associated with property are special and not subject to any kinds of restrictions. In fact, there is no important difference between regulations that may limit our use of land and those that affect the way we drive in traffic. Driving would be a completely chaotic experience if there were no traffic regulations. Traffic signs, stoplights, and speed limits allow us to drive more safely in congested areas. Adolescents (some of whom are no longer young) may believe that these regulations were created to thwart their desires, but most people recognize that the regulations are intended to ensure our physical safety. In an analogous manner, land use regulations that may prevent us from using our property in a particular way are not necessarily examples of the capricious actions of a dictatorial government. Instead, they usually are attempts to maintain order in society. Even though construction of homes and commercial buildings in areas subject to floods or earthquakes may produce short-term benefits to some individuals, ultimately, these activities impose excessive costs on the entire community, so restrictions on construction in hazardous areas make sense. No one has the right to impose costs on others by rejecting restrictions that provide benefits for the entire community.

Like it or not, if we wish to "develop" a flood plain, we have to impose some restrictions on that development to save individuals and the community from having to restore buildings damaged by floods. Those restrictions irk people who have plans for their property, regardless of the fact that the restrictions are intended to protect their investment.

Too often we have problems because we ignore evidence about why something is the way it is. We do not look at land as it is; we look at it as we wish it to be. That may be one reason many people reject evidence. It does not fit in with their desires.

But evidence represents reality.

SCIENTIFIC THEORIES

Anyone who attempts to talk about science to people who are not trained in one of its disciplines eventually hears the comment: "That's just a theory." That response is a convenient way to imply that the result is speculative, so there is no reason to believe it. That vernacular use of the word "theory" complicates discussions about science, because it muddies up the waters. The correct way to think about scientific theories is to recognize that they show us something about the real world. Let me repeat that.

A theory shows us something about the real world.

Consider this analogy. A theory is a window.

A scientific theory is analogous to a window because it provides a good, albeit restricted view of something. For example, the bedroom window of my house faces due west. From it I can see about one third of the property my wife and I own. I can see a pond that occupies about two acres, which exists because the previous owner had a dam built to trap drainage. Beyond the pond, is a barn, which we built for the sheep we used to raise, and beyond that is the building in which we operated a small business. In a sense, the view from the window hints at the commercial aspects of our life.

But the window provides no information about the twelve or so acres of pasture which are part of our property. If I walk to the other end of the house and look out the kitchen window, which faces due south, I see a few of the acres of the pasture that we own, and some property owned by a neighbor who has a horse grazing in his pasture. But I do not see our barn, nor do I see the bulk of our pastures. In a sense, this view shows part of the "bucolic" aspect of our life.

So the two windows provide restricted views of our property and our life. In some respects, the views are representative of what we own, but they do not show everything. Neither view shows the entire "package."

A scientific theory is like a window, in that it provides a clear view of some aspect of the natural world. It allows you to "see" what is relevant to it, and obscures what is not relevant to it. For example, Newton's theory of gravity explained most of the phenomena of interest to scientists in the 17th and 18th centuries. To mention just two, Newton was able to show why the orbits of the planets around the sun are elliptical and with the same theory (the same theory, not the same set of calculations), he was able to explain the cause of oceanic tides. But Newtonian physics could not provide information about the electric and magnetic phenomena which were studied in the 19th century. The work of Faraday, Ampere, Ohm, and Maxwell, culminating in what are today called Maxwell's equations provides a window showing how electromagnetic processes occur. And of course, the nuclear processes studied in the 20th century raised the shade on another window.

The point is that new theories were developed rather than trying to force Newton's theory of gravity to accommodate phenomena studied in later centuries. Even though the planets have magnetic fields associated with them, there is no point in trying to work that aspect into discussions of their orbits around the sun, because gravitation and electromagnetism do not interact in ways that affect the orbits.

So scientific theories are not "complete," in that no theory explains everything, but each separate window does explain enough to be worth using it to "look" at some aspect of the natural world. I can look out the bedroom window of my house and estimate the size of the pond. If I then go out and measure it in some way, the measurement will agree reasonably well with the estimate because the view from the window provides enough information to get a reasonably accurate value. That is what theories are for: to explain how things work and to allow us to make calculations.

A theory may oversimplify the phenomenon it describes, but it establishes ways to organize and evaluate information pertinent to it, and in doing so, provides ways to exclude information that is extraneous to one's analysis of the topic. So a theory focuses our attention, in that it defines what is relevant and what is not (for example, as mentioned, electromagnetic effects are not relevant to a study of the orbit of a planet). The theory may be incomplete but it still serves a purpose, in that it regulates a discussion of the subject in the same sense that the "rules of evidence" regulate what is admissible in a court of law.

Of course, because a theory oversimplifies the topic it describes, it is subject to revision or refutation by new information - as Newton's theory of gravity was subsumed by Einstein's general theory of relativity. That does not challenge the validity of Newton's theory: just because his theory of gravity does not explain everything that happen near very dense objects, such as stars, does not mean that we cannot use it to send a capsule to the Moon.

A theory is a useful tool, which is why I react negatively when someone says something is "just a theory," the implication being that it need not be taken seriously. That use of the word theory is invariably used by people with no background in science. In fact, it provides a useful criterion which distinguishes between scientists and those who know nothing about any aspect of science. After all, gravity is just a theory. But the theory of gravity works well enough that no one says it is "just a theory" in a derogatory manner. The theory is part of western culture because it explains so many phenomena that even people who have not taken a course in physics understand the consequences of stepping off the roof of a tall building.

Before going on to another example, I will mention a useful distinction. Gravity is a fact. Things fall down, not up because the earth exerts a force pulling them toward its center (which is downward). The theory of gravity explains how various phenomena which are affected by the force of gravity occur. In brief, *the theory explains how something which is a fact works.* Newton did not "invent" gravity: people knew that things fall down; Newton explained *why* they fall down. He came up with a mechanism which explained a fact that everyone was aware of. The distinction between a fact and a theory which explains the fact is important because it is not limited to physics. It also allows us to discuss evolution intelligently, something difficult to do when people dismiss it as "just a theory."

Many people think that the theory of evolution holds that humans evolved from apes, such as gorillas and chimpanzees, but it does not. In fact, the theory of evolution associated with the name Charles Darwin holds that humans and the contemporary apes evolved separately from a common ancestor by a process called natural selection. Understanding the process involves recognizing that as environments change, individuals more suited to the new ones are more likely to survive, reproduce and leave offspring, offspring which themselves will reproduce. Eventually, this selection process results in changes. It is true that the common ancestor did not resemble modern humans, but the branching out of organisms from a common ancestor explains why there are so many different *kinds* of animals. The transitions from that common ancestor to today's primates (including humans) are not known precisely because of the fact that for the last two million years at least two and as many as five species of primates (upright, walking, ape-like animals) existed simultaneously at any given time. The different species did not necessarily exist as neighbors, but the fact that sets of them lived at the same time complicates attempts to determine which of them continued to evolve to produce any one of the primates living today.

Without the theory of evolution, it is difficult to explain why two human-like species lived simultaneously during the last Ice Age - the Cro-Magnons and the Neanderthals. The Cro-Magnons are surely our ancestors, whereas questions about the origins of the Neanderthals have not been resolved. The skeletons of the Cro-Magnons cannot be distinguished from ours, except for the fact that on average they had slightly larger brains (perhaps they were smarter than we are). The Neanderthals were also upright walking primates, but their foreheads sloped inward, their jaws were more massive, and they "slouched" a bit because their spinal cords connected to their skulls a bit farther back than ours do. Attempts to sequence DNA molecules from Neanderthals have been only partly successful, but what has been done indicates that the two species were able to interbreed. Both should be called "human" because they both made tools and both had some form of a culture. The Cro-Magnons produced the paintings on the walls of caves in France and Spain, and some artwork dating back before the Cro-Magnons developed has been found, undoubtedly done by the Neanderthals, who also seem to have had burial rituals, in that colored pigments have been found on some of their skeletons.

The co-existence of two different subspecies of "humans" just a few tens of thousands of years ago should not be surprising: after all, two kinds of chimpanzees exist. And genetic evidence indicates that within the last 100,000 years two more species of human-like organisms (before the Cro-Magnons but along with the Neanderthals) lived in different parts of Asia. (That number does not include Homo erectus, which used to be called Java Man and Peking Man). If at least (or five) different human subspecies species did *not* evolve from a common ancestor sometime in the past, where did they come from? Explanations are simpler if we accept evolution as a fact. Just as Newton did not invent the idea that things fall down, Darwin did not invent the idea that evolution is the cause of the diversity of plant and animal life, but each of them was the first to provide a plausible mechanism: Newton, with the idea that objects exert gravitational forces on each other, and Darwin, with the idea of natural selection. Scientific theories work well, because they are not labeled theories until they can consistently explain some aspect of the natural world reasonably accurately. The word theory has no negative connotations for people trained in science; it is a useful aspect of our culture, because it provides the kind of vision that windows provide.

MORE ON FACTS AND EVIDENCE

In principle, there should be no problem in using the word "fact," but strictly speaking, the word can be used in a variety of ways. *The American Heritage Dictionary* gives four ways.

Something known with certainty.
Something asserted as certain.
Something that has been objectively verified.
Something having real, demonstrable existence.

The third version, something that has been objectively verified, is a good one to start with. In this context, "objectively" means that other people in other laboratories have done the test that I may have done, and obtained the same results. Someone may doubt my ability to do the experiment properly, or he may believe that I am biased, and so my results can be questioned. But those objections are harder to defend if several other people, who may not even know me, have obtained the same results.

So, if something has been objectively verified, it seems reasonable to invoke the fourth definition, namely that it is real. Of course, if attempts to verify a result are not successful, the consensus that develops will say that the original claim was wrong, and the results do not represent reality, and so they are not "factual."

An example of this process occurred in 1989 when two chemists at the University of Utah announced that they had been able to produce "Cold Fusion" by getting deuterium (a heavy form of hydrogen) atoms to fuse together at "room temperature" and release energy. If verified, the process promised to result in reactors that could produce large amounts of energy much more cheaply than from standard nuclear reactors or from coal-burning power plants.

A number of other scientists tried but were not able to duplicate their results, so the scientific community decided that the experimental procedures were flawed. Cold Fusion had not been obtained. It was not a fact in the sense of something known with certainty, and certainly was not one in the sense that it had been objectively verified.

On the other hand, there is no real doubt about climate change, because the third and fourth definitions of "fact" apply to it. Of the several thousand professionals who contend that the climate is changing, there is a lot of overlap, in that different people do various kinds of tests, or field measurements. The same kinds of tests and measurements are made in different parts of the world and on different aspects of the environment, so the criterion that studies have been objectively verified is satisfied. The results are known with certainty. The results can be considered to be reliable aspects of reality.

So it is safe to say that facts about global warming provide evidence that the climate is changing. That is, different kinds of facts support the contention that we are causing irreversible changes in the climate.

If you want to deny that the climate is changing, you have to find a way to deny the facts.

To be considered credible, it is not enough to say, "I don't believe the contention, regardless of the facts." Or, "I don't believe that the correlation between temperature and CO_2 is anything more than a coincidence." Or, that "temperatures have fluctuated before, so this one could be due to a natural occurrence."

It is not enough to say such things, that is, if you want people to think you are credible. That you are reasonable. That you do not get all of your information from sources with ties to the fossil fuel industry.

People who know something about climate science agree that the evidence supports the claim that the climate is changing. In fact, if we refine that last sentence a bit, and consider only the thousands of people professionally trained in scientific disciplines related to climate science, about 97% of those who do active research and publish on the subject agree that climate change is occurring. All but a few percent. To those cynics who say that those scientists say what they do because they need to continue to get grants to pay their salaries, I would respond that most of those who deny that climate change is occurring are funded by the fossil fuel industry. Several thousand professionally trained scientists working largely independently of each other, versus perhaps a dozen or so "deniers" who work at Conservative think tanks that are financed by the fossil fuel industry.

Which group is more credible?

Now consider the claim that human activities are responsible for global warming. The evidence that should convince anyone who is open to accepting evidence involves the studies of the forms (isotopes) of carbon found in carbon dioxide in the atmosphere (described in the chapter on Facts and Evidence), which indicate that the combustion of fossil fuels is responsible for the increase in carbon dioxide in the atmosphere since the early 19th century. That is an aspect of reality that can be taken to be factual because the data on the ratio of two isotopes of carbon are reliable. The "fact" is that human activities are responsible for the increase.

Facts are an aspect of reality, and they provide evidence to support a claim about something that is occurring. Scientific studies do not always explain everything we might wish, but without seriously studying a topic, questioning the validity or significance of these aspects is equivalent to questioning reality.

This is a good place to repeat something I said in the Introduction.

When science says that something has a basis in the laws of nature, it is saying that the "something" is an aspect of the real world. A reliable aspect. One which has been subjected to independent verification.

Turn that around and you get: something which has been subjected to independent verification is reliable. It is an aspect of reality. Consider it to be provisional until you can refute it or find something better, but don't reject it out of hand, because most of the time, you will be wrong.

DIFFERENT TAKES ON FACTS

Denying reality is a real challenge because you have to reject things (facts) that make sense. So you have to be nimble if you are going to be a "card-carrying" climate-change denier or creationist. The trick is to accept things that do not seem to matter, and deny the significance of those that seem to be important.

Keep in mind that I am not complaining that different people often have different opinions. I am saying that opinions do not always reflect reality, so one would hope that important policy decisions are based on more than opinions; they should reflect the most appropriate verifiable knowledge available. I concentrate on the examples of climate-change and creationism, because the mental gymnastics required to hold the positions involved with those "opinions" are tortuous, to the point of being impossible for a rational person to deal with.

For example, here some facts about climate change, and typical responses from skeptics.

Temperature measurements made since 1880 show that the planet has been warming. In addition, carbon dioxide concentrations in the atmosphere have been increasing since the start of the Industrial Revolution.

> The skeptics will usually accept these two facts, but respond with "So what? Temperatures have fluctuated in the past, and why should carbon dioxide be a problem? After all, it is essential for plant growth."

How about this fact? Carbon dioxide is a greenhouse gas, so it absorbs heat radiation emitted by the planet.

> The response to this fact is that the greenhouse effect is beneficial, because plants grow in greenhouses in the winter.

What about the relationship between temperature and CO_2? The correlation between carbon dioxide concentrations and the temperature of the atmosphere is statistically significant.

> This fact is usually discounted, partly because most people do not have a background in statistics so they do not understand the concept of statistical significance. And those who do understand statistics, just say a that correlation does not imply a causal relationship. They insist on *proof* that one causes the other.

Alright, let's talk about *proof.* Isotopic studies of the carbon dioxide in the atmosphere show that the major contributions of this gas are from combustion of fossil fuels, which can only be due to human activities, so the rise in temperature has to be due to the increase in carbon dioxide.

> If they have heard about this fact, and understand it, they will reject its relevance, but many of them will not

understand its significance because most of them know nothing about chemistry, and so they will not understand that this fact provides the "smoking gun" which proves that combustion of fossil fuels, and therefore, human activities, is causing the increase in carbon dioxide in the atmosphere.

What about the fact that numerous glaciers are melting back more rapidly than recorded in the past (indicating the effects of warming are felt in the Arctic and Antarctic zones)?

Glaciers do not impress them. They say that glaciers have always fluctuated, ignoring the fact that the glaciers are receding, not fluctuating.

Well, how about the fact that the warming is starting to change the living habits of numerous animal and plant species (indicating that climate change is occurring).

The effects on animal and plant species do not impress them. Who cares about rodents and butterflies and daisies? Animals migrate all the time, and plant seeds are spread by birds. So, shifts in breeding and growing zones are not important.

One last fact. In the summer of 2017, for the first time in history, cargo ships have been able to travel across the Arctic ocean without the need for icebreakers to accompany them.

The response to this fact is that "Opening the Northwest Passage will be good for commerce."

These are the kinds of responses that occur in media accounts of global warming and climate change. Basically, they are attempts to deflect attention from the fact that the "deniers" have no sound way to refute the conclusions drawn from the facts of the topic. They indicate either that the respondents know nothing about the subject, or that they are determined to deny the significance of the facts, at all costs.

Now consider the relevance of those responses in some detail. Start with the first two facts.

The average temperature of the planet has risen since systematic measurement began in the late 19th century (Measurements were taken before 1880, but not always at the same places, on a regular basis, so the reliable data set starts then). Over the same time interval, the concentration of carbon dioxide in the atmosphere has also risen abruptly. Those two facts are not usually denied because by themselves, they do not have political or economic implications.

However, the inference based on them, namely that the *cause* of the warming is the increase in the carbon dioxide, is different. The correlation between the two data sets is strong enough to be considered to be "statistically significant," meaning that there is only a small probability that the correspondence could be due to chance. But that inference is rejected by climate-change deniers, because the linkage between the two data sets supports the calls for major policy changes in the way energy is used to generate power to run the nation's economy.

The deniers content that there is no proof that the link between the two is real. Statistical significance does not *prove* a relationship, they say. They want some sort of physical or chemical arguments before they will concede a causal relationship.

So how do they deal with the isotopic studies of the carbon in the atmosphere, in Chapter 1, the studies that provide the requested proof? I have not seen any attempt to reject the studies, perhaps because there is no reasonable way to do so. No one has said that the studies are flawed. No one has mentioned any other studies that would indicate a different source for the carbon. If other kinds of studies would be relevant, why aren't they mentioned?

Now, what about the evidence from studies of animal and plant species? If their migrations are not indicative of climate change, why not tell us what are they do indicate? After all, animals and plants live in environments that are suited to them; ones which provide them with the nutrients they need; ones that are not too hot or too cold for them to survive in. Why would they migrate? What, other than changes in their environments, would cause species to migrate to higher altitudes or higher latitudes? And changing environments are what we are talking about.

And, finally, the sizes of glaciers are decreasing. The changes are monotonic, not fluctuating, so if they do not indicate climate change, what do they indicate?

Deny the significance of facts. Deny the implications of measurements and observations. Deny the results of studies that have been verified by independent observers – that is, ones that have been repeated by other people, who obtained the same results. That is what the deniers do, because it is all they can do.

What kind of world do these people live in? Do they have any idea of how natural systems work? Do they really think that the physical world works the way they want it to work? If they can cavalierly reject how the world around them works, just because they find something inconvenient, how much credence can you give to their comments about anything?

One more point should be considered. Many of the critics of global warming and climate change base their complaints on the fact that many of the predictions made by climate scientists are based on the use of computer models of the atmosphere-ocean system. The critics point out that the models do not represent perfectly all of the things going on, so their predictions cannot be accurate. But no one claims that the predictions are accurate, in the sense that the models can predict the weather in Chicago on June 15, 2055. That is not what the climate models do.

Let's starts with the fact that there are at last count, 25 different climate models involved in the studies summarized in the U.N. reports. They have been developed in different countries by different teams. They all solve the same sets of equations that govern how radiation, and the flow of heat and fluids interact, but the models differ in how they represent some of the fine details of the ocean-atmosphere interactions. Because of the differences, the models give slightly different answers to questions about sea-level rise and temperature increases in the future.

But the predictions of the models are consistent, in that they all predict future warming and future rises in sea-levels. So, although the absolute values predicted differ, and although the predictions made may not agree exactly with the measurements that are being made today, the fact that all of the models agree that certain things will happen, and the fact that those things are consistent with common sense, indicate that whatever oversimplifications exist, they do not produce important errors.

It is true that the models are oversimplified (no supercomputer is "super" enough to be able to consider everything that is happening), but the scientists who developed them are aware of what kind of things are represented in a simplified manner, and they are aware of what things are not considered at all. They also are aware of how those omissions are likely to affect the models' predictions. That is why the reports issued about climate change are expressed cautiously. But the critics are using the "oversimplification" complaint as a smokescreen, because the bulk of the recommendations in, for example, the U.N reports on climate change, are not based solely on the predictions of the models.

Let's be frank. The computer models are not capable of making precise predictions of climatic changes 50 to 100 years in the future. Precise, in the sense of accurate temperature values and inches of sea level rise. But that is not their purpose. The purpose of a computer model is to investigate how sensitive the climate is to each of the variables analyzed in the computer program. For example, if you are interested in what kinds of things will be most effective in reducing or eliminating climatic changes due to global warming, you might want to compare the relative effects of maintaining the rain-forest acreage and a reduction in carbon emissions from burning coal in power plants. The point is, that the models allow us to evaluate the relative effectiveness of different policies. The actual numbers that will be provided are not as important as the relative numbers, which indicate which policies are likely to be most effective. You don't need absolute accuracy for that.

The output of a computer model is intended to let us compare the benefits of different policies that might reduce the detrimental effects of continued global warming. As such, the models are useful tools, and anyone who criticizes all recommendations made by climate scientists because models have been used in parts of their studies, is using the models as a Trojan Horse to convince the public that all climate studies are worthless. They are wrong about that, because they know little to nothing about the models and how they are used. But the real problem is that their criticisms are intended to convince decision-makers that carbon emissions are not a significant danger to society

But carbon emissions, in fact, are a danger to society. To society, not to the planet. The planet will get along just fine if our activities are so destructive to the environment that our species becomes extinct. The planet does not need us: we need it.

So much for climate-change deniers. Now, what about creationists? Here are some facts which provide evidence to show why creationists should not be taken seriously when they claim that what they say involves science, followed by their typical responses to the facts.

Measurements of the amounts of Uranium-238 and Lead-206 in minerals in meteorites combined with the rate at which

Uranium-238 decays to Lead-206, indicate that the solar system, including the Earth, is billions of years old. The current estimate is 4.5 billion years.

> Creationists insist that the age estimates cannot be correct because the Bible indicates that the planet cannot be older than about 6,000 years. They use that age to argue that the theory of Evolution cannot be correct, because a few thousand years would not be sufficient for the diversity of living organisms we have today to have evolved.

The second fact is that the Sun does not revolve around the Earth, regardless of what is inferred in Joshua 10:12.

> Some biblical literalists have argued otherwise. In the 1970s, I subscribed to a periodical published by a group of Geocentrists. The articles insisted that the Earth is in the center of the solar system, because that is what numerous references in the Bible imply.

The next fact is that there is no archeological evidence supporting the claim that the events in the Book of Exodus occurred.

> Creationists have to reject the archeological evidence because the Exodus is an important part of the "history" of the Jewish people. They were not a "people" before it happened, and before the subsequent bond with God, represented by the gift of the Ten Commandments, which created their society. So, a number of attempts have been made to determine just where they crossed the Red Sea, and when, and who the Pharaoh was who pursued them. Nothing reliable has been found.

The last fact I will mention involves the Flood. The Noachian Flood could not have happened because the amount of water on the planet (in the oceans, glaciers, groundwater, and water vapor in the atmosphere) is far too little for the world ever to have been covered to the highest mountains by a planetary flood. Using some basic arithmetic, I demonstrate that statement in a subsequent page.

> Creationists need the story of the Flood to maintain a relationship between God and man. It is an attempt to stress the importance of maintaining their beliefs. Some attempts to document the existence of the Flood have involved claims that the "carving" out of the Grand Canyon was due to the Flood, contradicting more than a hundred years of geological studies of the feature.

Now let's consider these facts and the responses of the Creationists.

The age of the Earth has been determined by studies of the amounts of radioactive elements in minerals in very old rocks, and even better, in meteorites, which represent the left-over scraps that did not coalesce into planets when the solar system formed.

Creationists claim that the age determinations are flawed because they depend on the assumption that the rate of decay (or the "half-life") of a radioactive element has never changed in the past. They contend that since no one was around a billion years ago to measure the rate, we cannot be sure that it has never changed, and therefore the "date" obtained is not reliable.

The evidence against that argument is in the chapter titled "Making Assumptions," where I showed that the decay rate of Uranium-238 *cannot* have changed in the past. The argument presented there shows that the creationists' claim is unfounded. That "assumption" cannot be considered speculative: it is a fact, which destroys the only objection they can make about geologists' determination of the planet's age. The Earth is not "young," as biblical literalists claim.

Regarding the solar system, the comment in the book of Joshua that he commanded the Sun to stop moving across the sky, indicates that those who wrote the story believed that the Sun revolves around the Earth. That is not the case, and even the most ardent creationists today have to admit it. I understand that some people claim that what Joshua intended was to have the Earth stop rotating, but that is not what he said, and it is implausible, because Joshua could not have known that the Earth rotates. But the real problem is that Newton's First Law would require that if the planet stopped rotating, everything that is not tied down would continue moving – at several hundred miles an hour. Complete chaos would occur. You cannot cherry-pick what you insist is literally true and what is an interpretation. If the Bible is literally true, Joshua thought the Sun orbits the Earth.

Regarding the Exodus, archeologists have been all over the Sinai desert looking for evidence that several hundred thousand people spent forty years in the region, in the second millennium BCE. No evidence of such an occupation has been found.

The Exodus did not happen.

And finally, the story of the Flood. It is easy to show that a world-wide flood could not have occurred. A simple calculation shows how much more water would be needed to do what the story claims. Bear with me in what follows. Rather than using equations, I am doing it with words. You need to remember that the volume of a sphere is proportional to the cube of its radius. The formula is four-thirds times π, times the radius cubed.

I will start with the answer. The average depth of the oceans is approximately two miles, so to cover everything up to the tops of the highest mountains (e.g. Mt. Everest, at 29,000 ft. above sea level), would require about 87 times more water than is in the oceans now. Eighty-seven times more. That is not a trivial amount. Hurricane Harvey was a mild April shower compared to that deluge.

To see why that amount would be required, note that the peak of Mt. Everest is about 5.5 miles above sea level, so the depth of the water would have to be 2.8 times greater than at present (because 5.5 is 2.8 times greater than 2), so the radius of the spherical body of water must *increase* by a factor of 2.8 cubed, which is about 20. Multiplying that by four-thirds π gives 87. So, we would need about 87 times more water in the oceans today to accomplish the task described in the story of Noah and the Ark.

That leads to two interesting questions, namely, where did all that water come from, and where did it go when the waters receded?

If you object that those who wrote the story of the Flood did not know about Mt. Everest, I will respond that the source of the story is supposed to involve divine revelation, and surely God knew about it. If creationists insist on a literal interpretation of the Bible, they cannot cherry-pick which parts are flexible. But, if you must, by all means do the calculation for Mt Ararat. The numbers still don't work. You need about 17 times more water. Seventeen times more water. The same questions are relevant: where did the water come from and where did it go? The Flood could not have happened.

In order to dispute what is covered in this section, you must, in effect, say that everything we know about physics, the geometry of the solar system, and the ability to decipher archeological evidence is all wrong. So, disputing the facts presented here requires a vision of the world that is based on a book written in the seventh century BCE, which was partially based on mid-eastern folklore from the Sumerian civilization that existed two thousand years before (e.g. that is the source of the Flood story, which did not involve someone named Noah). To creationists, their source book takes precedence: nothing from modern science and mathematics that disagrees with the Bible can be trusted.

They don't believe that facts count. Facts that indicate that the planet is billions of years old are considered to be wrong. The simple calculation that shows that the Flood could not have occurred doesn't count. The geometry of the solar system doesn't count. The archeological evidence that shows that the Exodus could not have occurred doesn't count. Evidence doesn't count.

Facts and evidence are irrelevant if you are willing to accept only what is in an ancient book whose provenance indicates it is a collection of tales stitched together to bind a people together, not to describe how the world works.

Regardless of how fervent their religious beliefs are, they are denying reality.

HOW SHOULD SENSIBLE
PEOPLE RESPOND?

What should you do when someone says that your comments about the dangers of global warming and climate change are "political," and therefore need not be considered as "serious" comments?

First, you should point out, that the response is a distraction, because it is intended to make your claims, however strongly based on science, seem less credible, by implying that they are part of someone's agenda, and therefore cannot be valid scientific results.

Then you need to point out that no serious critic disputes a scientific result that way. Serious people know that the results of a study are based on the data collected and the analysis of the data, so any negative comments about a study that suggests policy changes have nothing to do with the validity of the study.

It is safe to assume that any accusation that a study is political is an attempt to divert the attention of other people from the conclusions of the study.

And, of course, accusing someone of making a claim political is itself a political statement, one which is intended to advance someone else's agenda. Unfortunately, that is too subtle for some people. And when scientists deny the accusation that the results are political, the response is that the accusations *would* be denied, because the scientists have a vested interest in the study. When people are determined to reject something, they will.

When attacked for being political, perhaps the best approach is to ask just what is going on. Why are these people responding this way to conclusions based on statements of facts? One way to do this is to examine the motives of a critic by doing what I recommended at the end of this book's Introduction, namely examine the "structure" of the criticism. Ask the critics to be specific: what was wrong in what was said? Don't just say it was political: tell us what was *factually incorrect*. Going on, ask if the data were collected appropriately, and analyzed appropriately. And do the people who criticize the results have the credentials needed to comment on the analysis?

A cogent response to these questions is needed to have a thoughtful discussion. But there never is one, because saying a conclusion is political is the tactic of someone who cannot evaluate the validity of the conclusion.

Consider this analogy. If someone gives you a noun, but not a verb, you cannot come up with the predicate of the sentence – presumably the conclusion - because it could be anything. In this context, if you ignore facts (nouns) and the logic pertaining to them, (the verb), you cannot determine if claims are believable.

For example, if nothing is factually incorrect, what is the problem? If you happen to work for an organization that is funded by the fossil fuel industry, you probably will oppose efforts to cut back on carbon emissions. So you will attack studies that disclose the effects on the planet's climate of burning coal. But do your objections mean that the studies you criticize *must* be wrong? Are you saying that even if the recommendation to curtail carbon emissions is based on factual evidence and sound thinking, you are still opposed to it? Why?

I am using the pronoun "you" here and in the next few paragraphs for convenience, not because I think that you, the reader, are a climate-change denier.

If nothing about the results can be *proved* to be incorrect, why would someone object to using the results to develop a policy? That kind of objection is political, not scientific (Both sides of this topic can use that ploy). There is a difference between just being opposed to something and claiming that it is wrong in the sense of being factually incorrect. Does a correct statement become wrong just because someone whose view you disagree with likes it?

If something switches from right to wrong just because you disagree with those who think it is right, what does that say about your motives? Or about your grasp of reality? If you are a Republican, and a Democrat says the Sun will rise in the East in the morning, will you disagree with the statement? Under what conditions will you disagree with that or any other fact that has been verified by independent observers?

Now let's increase the resolution of the discussion. Do you have a problem with the data set? For example, was the collection free of bias? Were the data points collected independently of each other? Were the measurements sufficiently accurate? If you cannot show that there are problems with the numbers, what is your objection?

Going even further, was the analysis of the data appropriate? That is, was the correlation coefficient appropriate for the kind of data collected? Was the significance table used to evaluate the strength of the correlation appropriate for the correlation coefficient used?

Tell us what your problem is.

People who criticize a study, should be able to answer all of those questions, so that brings up the next point, which is the credibility of the critics.

Have the critics of the study provided any answers to the questions just mentioned? Do they even understand the questions? Have they read the study carefully, and do they have the background necessary to understand the details of the study?

To be frank, the climate-change deniers rarely address any of the kinds of questions asked here. They just say that there is disagreement about the cause of what is observed, without pointing out that the disagreement is manufactured by people who do not want to admit they are wrong. When they really are not qualified to make sensible comments, they may resort to saying that the study represents "junk science," hoping that their audience will not ask how they know it is junk.

This line of questioning could go on and on. For example, how do the critics define "junk" science? Don't use the climate studies as examples; give a general definition and other examples. If they can't be that specific, why should anyone take their complaints seriously?

Unfortunately, the news media do not ask these kinds of questions when politicians and climate-change deniers dispute scientific studies.

George Lakoff is a social scientist who has written extensively about how people hide behind mental barriers to justify their approach to life. He refers to mental structures he calls "frames" as their basic tool, a frame being a structure defining their views. For example, a "patriarchal" frame is used by people, nearly always men, who are authoritative by nature, and who feel that a vertically structured, ladder-like, hierarchical society is the "natural" one. Of course, they also feel that men are at the topmost rung on this "ladder." When you are on top, it is easy to criticize everyone and everything else.

Lakoff contends that if you want to convince someone that his frame is not justified, or even that it is repellant, it is not enough to criticize it or even to show that it is invalid. You must have an alternate frame to replace it. Simply "refuting" a frame just tends to reinforce it. For example, consider Conservative politicians' favorite economic frame, which for simplicity, I will call "Trickle-Down."

The basic idea is that anything government does to benefit those in the topmost component of the income distribution will benefit everyone in society. Lowering taxes on the wealthy is the favorite example of those for whom this frame is the only appropriate way to think about economics. Their argument is that lowering taxes puts more money in the hands of the public, and when that money is spent, the economy is stimulated. It follows, they claim, that because those in the upper-income brackets pay the most in taxes, they should get the largest tax breaks, in which case, the effect on the economy will be magnified.

There is no point in telling people whose life is structured by this frame that a study showed that there is no significant correlation between cuts in the highest tax brackets and the Gross Domestic Product (which is a measure of the strength of the economy). They will not relinquish a belief that makes so much sense to them, one they have held for their entire lives.

What is to be done? Lakoff recommends that instead of showing that a frame is wrong, one must provide an alternative, one that no one can deny. Using one that shows the consequences of an inappropriate frame is a good approach. One that shows what it does to the person you are speaking to. One that shows how it hurts him or her. In fact, the study relating tax cuts to the G.D.P. showed that the result of tax cuts was an increase in income and asset inequality, because the bulk of the money went to and stayed in the pockets of those in the uppermost one percent of the income distribution. And they park it in offshore accounts that do nothing for the economy or for the bulk of the population.

One wonders why average people are willing to be subjected to a policy that hurts them in order to benefit the very wealthy. The average person may not be able to answer such a question easily, but one hopes that they would try. Cognitive dissonance may not be good for the digestion, but it is the only way to get people to discard erroneous frames.

Another example could be climate change.

There is a scientific consensus on global warming and climate change. The consensus position is that the planet is warming and that the cause of the warming involves human activities, primarily the use of fossil fuels to generate electricity and to power internal combustion engines. Curtailing our use of fossil fuels will have significant effects on the economies in the developed countries, so resistance to doing anything about the problem of "warming" has developed, by groups funded by the industries affected. Because most people are not aware of the difference between weather and climate, it is easy to convince them that nothing important is happening, and so nothing drastic should be done to damage the nation's economy. The relevant "frame" provided by the fossil fuel industry is something to the effect: "Don't worry, because nothing important is happening now, and will not happen in the near future."

Merely telling people that some drastic changes in society are needed will have no effect. The only way to penetrate the bubble in which many people live is to show them how "going on our merry way" will result in conditions they would not normally choose.

We need to explain why our dependence on coal-burning power plants to generate electricity is going to result in conditions that will be detrimental to them. One way to do that by addressing three questions involving facts and logic.

What is happening and why is it happening? This involves Facts.
Why does it matter? This also involves Facts.
What should we conclude? This involves Logic.

Pause here for a moment and look at the last three lines. Many people refuse to accept facts that they consider unpleasant, but how many will admit that they reject logical arguments? Who will admit to being irrational?

First Question – What is happening,
and why is it happening?

The planet is getting warmer because of the extensive use of fossil fuels used to "power" the world's economy. Combustion of hydrocarbons such as coal and petroleum releases carbon dioxide, a greenhouse gas, into the atmosphere, thereby increasing the ability of the atmosphere to retain the heat emitted by the planet and thereby increase its average temperature. To understand why, think of the gases in terms of a blanket wrapped around the planet. As we add more of these gases, the blanket gets thicker, and keeps more heat close to the planet's surface.

Everything said in the last four sentences is either a fact or an inference based on physical laws. For example, burning coal, which is essentially pure carbon, produces carbon dioxide (CO_2). That is a basic fact from chemistry. No technological development is going to allow us to burn coal without producing the gas. And, the properties of the CO_2 molecule explain why it traps heat. No other explanation for the observed rise in temperature associated with the increase in CO_2 makes any sense.

Anyone who tells you that burning fossil fuels does not increase the amount of a heat-trapping gas in the atmosphere knows nothing about the topic.

The increase in temperature is causing changes in climate, which will result in significant changes in the ability of society to function as it does at present. Resistance to reducing carbon emissions to the atmosphere comes primarily from the fossil fuel industry, which contends that the problem is exaggerated, so no major changes in the way society functions are needed. The public relations efforts of that industry have been successful in convincing large segments of the public that even if global warming *were* occurring (and the industry would rather have us believe it is not), the steps needed to prevent it are so drastic that we need to wait until we are sure that there are no other solutions before curtailing the use of fossil fuels.

These public relations efforts have been successful because the public has at best a murky understanding of climate, as opposed to weather. The distinction between long-term and short-term fluctuations is not clear to many people, but understanding it is essential if long-term disaster is to be avoided.

Pause here for a bit. The resistance to taking steps to dealing with global warming comes from the fossil fuel industry. That is understandable, because if we stopped burning coal and petroleum, and left those compounds in the ground, the companies in the that industry would be bankrupt. Their "worth" depends on being able to extract coal and oil to be used to generate electricity and power automobiles.

It is ironic that people who suspect that the scientists documenting climate change are making things up so that they will continue to get government grants to support their work, do not seem to realize that everything they hear about the bias in scientific studies comes from an industry that depends for its very existence on ignoring the scientific studies.

Second Question – Why does it matter?

To repeat something said earlier, it is not enough to tell the public that those who deny that the problem is important are wrong.

Without an understanding of what is happening and why it is happening, the public cannot make an informed judgment about what needs to be done. It is necessary is to get the public to understand why society needs to curtail the burning of fossil fuels, and one way to do that may be to emphasize the difference between weather and climate – a distinction that is not clear to many people.

To provide an explanation, it is useful to start by thinking about different major climatic zones today, which I will call Tropical versus Temperate versus Arctic, each of which has its own short-term temperature variations of the kind we call "weather." Those variations differ in the different climatic zones, but they recur in each of them every year, which is the cause of the "seasons."

However, it is the long-term characteristics of each zone that we associate with "climate." Those characteristics include the long-term average temperature, the maximum fluctuations in temperature, and the phenomena associated with those fluctuations (for example, the existence or nonexistence of frost and snow).

It is useful to think of the long-term characteristics as large-scale versions of the "growing" zones that are familiar to gardeners. These zones specify which plants will survive in which parts of the country – because of the local climatic conditions. So, we assign the word "weather" to the daily short-term fluctuations and "seasons" to the existence of regular patterns," whereas "climate" *refers to the very long-term regularity of the seasonal patterns.* And "long-term" can mean decades or centuries.

If the planet is warming, both weather patterns and climatic zones will be affected, but in different ways. The simplest example for how warming affects weather patterns is that in a warmer atmosphere, the increased energy levels will result in more extreme weather events such as storms, hurricanes, and droughts. With respect to climate, we can expect every zone of the planet to be subject to higher temperatures in the future, which means that the warm climatic zones will expand laterally and encompass more land, so the tropical zones will expand northward (in the Northern hemisphere), the temperate zones will shift to higher latitudes, and the arctic zones will decrease in size, because even in the higher latitudes, the temperatures will rise.

Repeating the ideas in the second half of the last paragraph (which starts with "With respect to climate..."), as tropical zones get warmer, their northern "edges" (in the northern hemisphere) merge with the temperate zones. So the tropical zones grow laterally, and the area they cover increases northward. Similarly, the northern edges of the temperate zones get warmer, and merge with the arctic zones, which decrease in size (because they cannot grow past the North Pole).

Look at the last paragraph. Everything in it is a fact. Now let's consider some inferences based on those facts and basic common sense. Keep growing zones in mind.

As the climate changes, so will what you can grow in your garden. Will the "deniers" start to believe in global warming and climate-change when their flower gardens no longer bloom in the Spring because what they usually plant will no longer grow there? There will be dramatic, irreversible changes. Every zone will become warmer, but those that are currently warm will become much warmer, and some of them may become uninhabitable. In the mid-latitude regions such as the American Mid-West, which is the nation's "breadbasket," the change in climate will produce significant changes in what can be grown there. An increase of several degrees may not seem catastrophic, but when wheat can no longer be grown in the Midwest, the nation's agricultural operations, and therefore the nation's economy will be affected.

Are irreversibly harmful (yes, I said irreversibly harmful) changes to the economy because of higher temperatures enough of a reason to start doing something about global warming? Will that question penetrate the bubble in which many people live?

Third Question – What should we conclude?

The discussion in the preceding paragraphs illustrates why global warming is a topic everyone should be concerned about. Dealing with the problems caused by warming will not be straightforward, because it will require curtailing the use of fossil fuels to "power" our economy, which will have significant costs associated with it. But those costs are monetary (and what will money be worth when there is no food to buy?), whereas the costs associated with doing nothing to prevent climate change involve more than the monetary costs that Nicolas Stern estimated (which was discussed in the chapter on "Assumptions"). They involve serious, irreversible social changes, such as mass migrations of people on a scale never before observed, and the transformation of parts of the country (e.g. the American Southwest) into uninhabitable zones. How does one put a cost on those changes?

I cannot repeat often enough that every single thing mentioned in this part of the chapter is either a fact, or an inference based on well-understood physical laws. The only thing climate-change deniers can honestly quibble about involves the time-scale involved. No one can predict precisely how long it will take some of the dire consequences predicted to occur, but that is no reason to put off dealing with the problem.

It is dangerous to wait, because the atmosphere-ocean system is complex, and complex systems tend to have "tipping points," at which behavior changes radically and irreversibly. Irreversibly is the key word.

But the most important reason to do something is that to delay taking action would be to admit that one's grandchildren are not important.

POST-TRUTH AMERICA

What, you may wonder, does "post-truth" mean? It refers to the contention that "truth" is an old-fashioned criterion when one is dealing with important topics. Some people feel that it is no longer necessary to examine facts or evidence to determine what is true and what is not true. They feel that the appropriate response to contentions claims is to examine the motives of the people trying to argue using facts.

Of course, this is the philosophy behind most political arguments, but it also is applied to many discussions about scientific studies, and unfortunately, provides an excuse to ignore facts and evidence.

For example, the scientific consensus for global warming and climate change is clear, but in order to justify ignoring the ramifications of the consensus, its political opponents take its very existence as proof of a conspiracy. After all, why else would people with so widely different backgrounds, such as scientists and liberal politicians, agree with each other about the matter if they weren't engaged in a conspiracy to put something over on the public?

Note that this kind of response is from *political* opponents, not *scientific* opponents. What are they opposing?

They are opposing the consensus developed by a few thousand professionally trained climate scientists which holds that the rapid rise in the carbon dioxide levels in the atmosphere is causing global warming; that it is causing acidification of the oceans; that it is causing glaciers to melt in Antarctica, Greenland, and in the Andes and Himalaya Mountain ranges; and that it is melting the ice cover in the Arctic Ocean – among other things.

In addition to ignoring facts and evidence, the "post-truth" crowd is ignoring logic, because they will not admit that nothing else could cause all of these phenomena at the same time. So of course, the opposition to the consensus is political: it could not possibly be scientific.

The same points apply to creationists, who reject anything that contradicts their interpretation of the Old Testament. Their need to accept everything in that source as literally true supersedes any scientific results, so they have no trouble ignoring facts and evidence.

To political opponents, the fact that a consensus exists does not mean that the scientific community's judgement is most likely correct. To them, it is immaterial that several thousand professionals agree because the evidence is so strong that disagreement makes no sense. These people don't even consider such things. That would require some thought. It might even require them to reconsider their thoughts about the subject. But we can't have that, can we?

No, to these people, the way to deal with a consensus that they do not want to accept is to attribute it to a conspiracy. It is easier to claim that a conspiracy exists than to admit that they may be wrong, because politicians never want to admit that they are wrong about anything. And the fact that they know nothing about a subject makes it easier to ignore the consensus they disagree with. The same is true for climate-change deniers and creationists. Does that sound illogical? It certainly is, but we are dealing with politicians, climate-change deniers, and biblical literalists, so logic is not a consideration.

Politicians know nothing about climate change, so they cannot say why the consensus is wrong. Instead, they resort to saying that because some scientists disagree with the consensus, it makes no sense to do anything drastic until the science is sound. They ignore what a consensus means. The fact that a consensus exists implies that the science *is* sound, so the few scientists who disagree with it are most likely wrong. If you wonder why the opinions of a few scientists who disagree with the consensus should mean more than the opinions of the several thousand who contributed to the consensus, congratulations: you are using basic logic. To add another nail to the coffin, when I mention scientists in this context, I am referring to professionally trained climate scientists, not people with degrees in Civil Engineering or Computer Science or Food Science or whatever. I am concentrating on a large number of people who know what they are talking about. It is safe to assume that the conclusions drawn from their studies are correct.

As mentioned in the Introduction, a comment made by a sportswriter named Grantland Rice, back in the 1920s applies to the position taken by climate-change deniers. Borrowing a phrase from *Ecclesiastes*, he said:

"The race does not always go to the swift,
Nor the fight to the strong.
But that's the way to bet."

In other words, when you bet on a long shot, you usually lose your money. Unfortunately, in the case of climate change, much more than money is at stake.

Another ploy used by the post-truth crowd to deny reality involves distraction, by which they try to convince people that an elite group is telling ordinary people what to think. That distraction enables the widespread belief that everyone's opinion is equally valid, regardless of one's credentials, and provides "cover" for people when they reject various forms of knowledge. The opinion of some guy you overheard in a bar last night clearly is just as credible as the opinions of thousands of professional scientists who have been studying a topic for decades. Does that make sense? I hope not, because that is the subtext of the arguments used by these people. It is wrong and it is not logical, but that is all they have.

Moving right along, let's look at another way they deny reality. When scientific results are used by one side of a debate to justify its position, and the other side is determined to disagree, but cannot show that the results are wrong, that side usually tries to deny the significance of the results. And one way to deny the significance is to convince people that the people doing the study are biased; that they have some ulterior motive – often financial. Rumors are spread saying that the scientists need to keep getting grants to make sure that their salaries will be paid, even though most of them are academics with tenure. Do you know what "tenure" is? It is the academic equivalent of a "no-cut" contract. So much for that approach.

That kind of distraction works both ways, of course, because, in the case of climate-change, the majority of the "deniers" are not professional climate scientists. They work at Conservative think tanks, and they are paid from grants to the organization that are usually provided by the fossil fuel industry. When you work at a think tank, you are responsible for your own salary, so you have to make sure those grants keep coming in. But of course, no one at the think tanks feels that that is a conflict of interest. No one would suggest that these guys are just "hired guns," who would switch to the other side of the issue if that is where the money were coming from. That would indicate that these guys are as venal as they claim the scientists are. Suggesting that would be crass.

Why is it so easy for people to deny knowledge? Whatever happened to the goal of improving the public's scientific literacy?

Survey after survey show that scientific literacy is low in America, which should be surprising because so much money was spent to improve the teaching of science and mathematics after the launch of the Sputnik satellite by the Soviet Union in 1957. New science curricula were developed and programs were created to improve the education of science teachers. Yet positive results have not been obvious.

But why should they be obvious? The efforts were all aimed at academic courses, but how many adolescents take school seriously? Consider the fact that for generations, every student has also been exposed to Social Studies (Civics) courses, yet many adults confuse the Declaration of Independence with the Constitution, and even more have no idea what is in the Constitution, or for example, know that it was not until 1920 that women were allowed to vote in national elections. Why should academic programs aimed at adolescents turn out a generation of scientifically literate citizens?

But the problems discussed in this book are not directly related to a lack of scientific literacy. After all, there are many people trained in a variety of scientific disciplines who find it easy to deny the linkage between human activities and global warming. With the exception of a few of them, they are not climate scientists, but they have studied different scientific and engineering disciplines and many of them work in one of them, so they feel competent to have what they feel is an informed opinion on climate change.

So what is the problem? Why can't they see the forest for the trees?

The answer seems to be that people believe what they want to believe. And the things they want to believe are not sitting around in their memories independently of all of the others. A useful metaphor is a box in which opinions on a variety of topics are stored. The topics might include climate change, gun control, the evils of big government, national security, abortion, birth control, people on welfare, standing for the national anthem, the income tax system, etc., etc.

When you reach into the box to get one out, you find that others come too, because they seem to be stuck together. For example, if you pull out an article on gun control, something about "big government" is likely to come along with it. If the "income tax system" is pulled out, "people on welfare" may emerge with it.

The point is that when talking to someone about nearly any topic, you may not be able to address just that one, because you will find that the conversation shifts to topics in that person's mind that are associated with the one you want to address. Like Hercules trying to kill the hydra, if you manage to vanquish one, others quickly appear, and there may seem to be no way to get back to the original topic.

When people have strong opinions about a topic, it is not easy to get them to admit that they may be wrong about the subject. The linkages between what they thought and the other things in the box may be too strong to sever by just using logic.

Because our educational system is based on industrial management principles in which considerable time is spent preparing for standardized tests which measure achievement solely by scores on those tests, there is not enough time to let students use what they learn. So, students cannot think about the topics covered and recognize when they are or are not related to other topics.

The result is that we turn out large numbers of graduates who drag a lot of baggage around. George Lakoff's recommendation about providing alternate "frames" does not always work, so we should not be surprised that so many people can casually disregard facts that underlie reality.

It should be obvious that the arguments used by the "post-truth" people do not differ much from those of the Relativists mentioned at the start of this book. To these kinds of people, "truth" depends factors that may have no correspondence with reality. They don't realize that if you are not willing to justify the details of claims that you make, you are not contributing to a fruitful discussion of whatever topic is presented.

People want to believe in magic. Post-truth may be magical, but truth is not.

Remember what I said earlier:

Truth is partial to knowledge. To deny truth is to deny knowledge, and knowledge is our link to reality. So, to deny knowledge is to deny reality.

ABOUT THE AUTHOR

Pascal de Caprariis has a B.S. in Geology and an M.S. in Geophysics, both from Boston College. His Ph.D. degree in Geology is from Rensselaer Polytechnic Institute. While in graduate school, he taught high school Physics and Earth Science in New York State before moving to Indiana where he taught Geology at Indiana University-Purdue University at Indianapolis until he retired.

Printed in the United States
By Bookmasters